Reaching Escape Velocity

Launching gonzo engineering projects
with sponsors, media, volunteers,
and other potent forces

by Steven K. Roberts

Nomadic Research Labs
microship.com

Published by

Nomadic Research Labs

http://microship.com

Library of Congress Control Number: 2009904079

ISBN: 978-1-929470-05-1

Printed in the United States of America

For bulk purchases, please contact *wordy@microship.com*

Cover photo by Steven K. Roberts: *Nomadness* at anchor in Utsalady Bay, Camano Island, Washington (2008)

Contents

Foreword

A Grand Vision is only the beginning. No matter how much passion you bring to bear on the project of your dreams, the odds of actually escaping the "gravity well" are low... unless you find a way to leverage larger forces. This document, derived from 25 years of audacious feats of gonzo engineering, presents the keys to six tools that are essential to a large-scale project:

I have contemplated publishing a book on this subject for years, and only now (2009) have decided to do so. It can be considered the collection of "trade secrets" that have made my adventures possible... the art of working with sponsors, media, and volunteers to get an insanely ambitious project off the ground and keep it moving on its own momentum.

This book has a specific audience: those who are attempting to "reach escape velocity" with a massive feat of engineering. It is not about hardware or software technical methods; it is about the meta-hack of developing enough support and buzz to get your project to take on a life of its own. Large corporations can do this with brute-force methods (unlimited money and people), but individuals face daunting hurdles when competing for mindshare and resources. Without the ability to leverage larger forces as a sort of "martial art," it is exceedingly difficult for a lone geek to escape the gravity well.

My own strange career has been the proving ground for the techniques revealed here. In 1983 I took off across the US on a computerized recumbent bicycle, freelance-writing and consulting while underway. This was bizarre at the time, though it would now be unremarkable. But the project took on a life of its own, and I eventually covered 17,000 miles on three versions of this increasingly geeky machine. With a handlebar chord keyboard, heads-up display and head mouse, console Macintosh, and 24/7 satellite net connection (in 1990) it was a sort of geek extravaganza... and the project had about 150 corporate sponsors, 45 volunteers, and almost continuous mainstream media coverage. It became self-supporting, then segued through the '90s into the Microship project—an amphibian pedal/solar/sail micro-trimaran with similar technological overlays.

This little book is not about any of that, but what it does do is explore the methods I used to pull the whole thing off without having a job or money in the bank. In these days of faltering economy, such techniques are more relevant than ever.

This is followed by a brief introduction to *Gonzo Engineering* that further explores the kind of lean, intuitive, art-and-engineering thinking that is necessary to proceed rapidly despite scarce resources.

As I said, this publication is not for everyone - mainstream R&D has little room for such social engineering, and if you're just carving out a niche as a freelancer, most of this is a bit over the top. The readers who will benefit from this information are those who are trying to pull off the impossible with an insanely audacious project, do an end run around traditional linear approaches to product design, or explore that strange territory best summarized by my old friend David Berkstresser in his immortal observation: *There's glory in using inappropriate tools. You can tell you're pushing a new frontier when all available tools are inappropriate.*

If any of that sounds deeply familiar, this will pay for itself very quickly.

The console of the Winnebiko II, *the second version of a computerized recumbent bicycle that covered 17,000 miles (1983-1991).*

Introduction

One day in retrospect
the years of struggle will strike you as the most beautiful.

- Sigmund Freud

It is perhaps the most-asked question I heard during a quarter-century of playing for a living: "How the heck do you *pay* for all this?"

Let me generalize that a bit. So you have a grand vision, a passion, a crazy dream of adventure. What does it take to start something this huge if you're not rich and surrounded by a team of the ultimate labor-saving devices (other human beings)?

Seriously, how the hell is one person supposed to begin something so fiendishly complex that it can sometimes take a book-length manuscript just to *state the problem*? An enduring life's passion isn't a weekend hobby project, or even a well-defined engineering job; it's overwhelming. In my case, the dream of travel and adventure was only part of it. I also had to take on the design and fabrication of an amphibian pedal/solar/sail micro-trimaran, launching me into the realms of composite fabrication, mysterious chemicals, custom machining, hydraulics, harsh-environment electronics packaging, networked embedded controllers, power systems, thermal challenges, marlinspike seamanship, antenna interactions, usability, survival tools, and thousands of lines of code... yet I'm an unemployed guy with no money in the bank, picking up a few bucks occasionally from articles, speaking gigs, and eBay sales.

Doesn't this seem just a wee bit unrealistic? Have you ever entertained a Grand Idea, then abandoned it with a shrug after contemplating the Herculean scope of the undertaking?

This is the point, right here, where most interesting potential projects stumble to a halt. There are a lot of creative people out there (if you're reading this, you are likely to be one of them), and as such there is no shortage of good ideas. But most of them never get past the "one of these days" stage, stymied before they even get off the ground by practicalities, daunting costs, and resource limitations. I've seen quite a few cobwebsites about way-cool machines that never made it past a fantasy and some exuberant HTML.

If you find yourself in this position with *your* dream project, this little document might turn out to be pivotal. I'm going to reveal all my "trade secrets," things I've never publicly discussed before—the body of techniques that have let me make a career out of playing with cool toys, traveling around, and hanging out with interesting people. Although I'll be referencing my own projects (microship.com) for context, this stuff is applicable to any large-scale, passionate pursuit. (Some of it might even applicable to reasonably-scaled, relaxed pursuits, but I have never had occasion to test that.)

The first problem is reaching escape velocity by injecting enough energy into the project to make it take on a life of its own and move beyond the daydream phase. I suppose you could throw a lot of money at it to achieve this end, but as I have no experience with that approach, I am instead going to tell you how to do it in true hacker style: circumvent limitations and optimize resources by manipulating natural forces and cultivating win-win situations, in the process expending as little effort as possible. (Yes, even in a depressed economy.)

We will discuss six key components: an overarching business structure, diverse educational resources, sponsors, media coverage, a stable audience, and a team of assistants. I'd better warn you now: some of this stuff involves making your life public almost to the level of exhibitionism, and that can be challenging for the terminally shy. But the audacity factor is critical and worth cultivating; it can pay off in amazing ways.

Facing Page: two more machines that emerged from my life of gonzo engineering. At top is the infamous 580-pound, 105-speed BEHEMOTH, *with Mac, SPARC, and DOS environments as well as satellite datacomm, HF/VHF/UHF ham radio, heads-up display, head mouse, handlebar keyboard, 6-level security system, speech synthesis, 72 watt solar array, and deployable landing gear to keep the monster upright on killer hills. The bike now resides in The Computer History Museum.*

Below is the Microship, *the result of an 8-year development project involving extensive sponsorship, students, and volunteer teams. This is an amphibian pedal/solar/sail micro-trimaran with retractible wheels, hydraulic systems, 480 watts of peak-power-tracked solar panels, and zippy performance under sail.*

Both of these, as well as the later Nomadness *project, are documented at microship.com.*

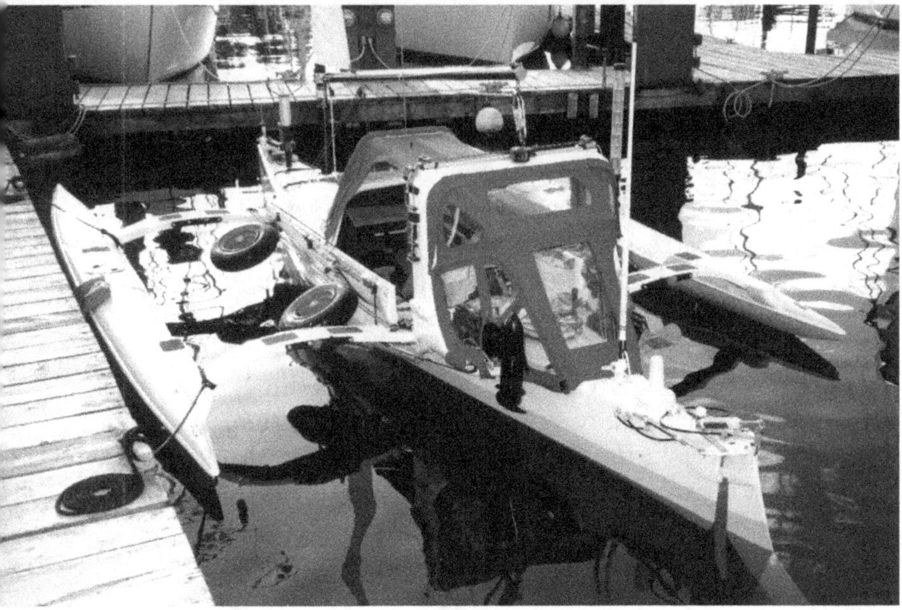

Chapter 1

The Business Angle

Before your eyes glaze over at the word "business," let me hastily say that there is nothing in here about setting up a company! Lots of people can tell you more about that than I can, and although this discussion covers a lot of territory, you won't find me talking about how to run a corporation, nor even a sole proprietorship with shoebox accounting. I find that stuff hopelessly confusing, and am thus perennially at the mercy of people who are comfortable with financial and legal matters.

Still, you can use the system to your advantage (even *without* running a shell game, which I don't recommend... stress and creativity don't mix very well). An obvious requirement of managing a huge under-funded project is to minimize your costs, and the first step in that direction is to create a *bona fide* business entity encompassing the field in which you are working—an enterprise that actually shows cash flow and can withstand an audit. In my case, since I don't want to be a boat builder, I have to come up with something (besides manufacturing Microship clones) that renders all project expenses honestly tax deductible. Fortunately, I've been a writer for a long time, and with an endless stream of articles and the occasional book about these projects, they can legitimately be defined as "research." The only tax-law requirement is that I show a profit occasionally, and somehow, bad work habits notwithstanding, I manage to do so now and then (with a little help from an online store, affiliate links, and spin-off nickel generators).

I want to give you a couple of basic ideas that can help cover the costs while wrapping your pet obsession in a cloak of business legitimacy.

It has become axiomatic that a key component of a large independent project is "buy-in" by the public (and, as we shall see shortly, a volunteer community). This has certainly been the case with the Microship and its land-based predecessors, but you also see it in open-source development communities such as those hosted on Sourceforge: there is a sort of "critical mass of publicity" that has to be achieved for a project to take on a life of its own. One of the best ways to accomplish this is to build a related business based on information and spin-off dissemination... magazine writing, self-published technical monographs, system documentation, shareware authoring, project kits, or anything else that leverages the work you actually *want* to do in order to simultaneously generate income and build exposure. There are very few professions in which you get paid to advertise, but that's basically what publishing is all about—the more you do, the more you create brand recognition and demand (assuming you're reasonably good at it). Magazine articles can lead to columns, which in turn can lead to books, which eventually, if you're lucky, can lead to a lucrative speaking career and a royalty stream.

Alas, many technical people either can't write very well or just don't want to; an old wag once said that writing is accomplished by staring at a blank page until little drops of blood appear on your forehead, and I know a lot of brilliant geeks who just don't have the patience or skills to make a go of it. But there are other means to the same end. Let's talk about consulting.

The best generalization I can give you is that the boundaries between specialties are where it's at. It is no accident that most projects in the domain of gonzo engineering are, by their nature, comprised more of new ways of combining existing technologies than of linear envelope-pushing; the latter, while honorable and necessary for ongoing industrial progress, is less likely to yield the kinds of breakthroughs that make the media flock to your door. It's not that there's anything wrong with it, it's just that individuals have a much harder time with "straight ahead" advances in the state of the art than do well-funded companies... that sort of work *does* lend itself well to structured engineering methods and thus tends to be the most likely course of corporate product development (think Moore's Law).

But an obsessed individual making leaps of intuition in the middle of the night is almost inevitably looking at new interactions between existing ideas—making novel connections across great chasms (like finding a mathematical hack to avoid the rotation and scaling problem associated with image recognition, based on the logarithmic polar mapping discovered through end-to-end electron microscopy of the optic nerve). So it is likely that a gonzo engineering project already has some of this cross-boundary action happening. Why is this relevant?

Because the hottest consulting action is almost never within your own specialty; it's when you take your accumulated knowledge, cross over to a client who speaks a different language in an unrelated field, and solve Major

Problems. If you know a lot about spread-spectrum data transmission, do you try to sell yourself to radio manufacturers that already have cubicles full of experts? Or do you become the hero of the hour to some company that has run into a wall trying to move data between ceramic foundries, deploying a few off-the-shelf radios and beam antennas along with a nice fat invoice for your trouble?

Note the parallel here. The best consulting opportunities feel a lot like a gonzo engineering project, and, if you're clever, can even use the same components and tools. So at the beginning of the mad quest to build your system, whether it's a Microship or the world's first standalone *Refrangible Densiosity Enhancement Device*, try to associate a business model with the project that can realistically turn all your R&D into a valid business expense and let you depreciate capital expenditures. If the coupling between the business and the project is close enough to stand up to the scrutiny of a bean counter, you not only save a ton of money but also create an aura of professionalism that will pay off again and again. Bankers, insurance companies, vendors, and even the trade press will take you much more seriously if that contraption you're building looks like part of a business instead of a "hobby."

Only *you* have to know the truth about the underlying motives... and won't you be surprised when the business angle is self-fulfilling and you really DO find yourself making a living from what you love?

Chapter 2

Getting Educated

Once that tiresome money-making stuff is out of the way, we can start to play. One of my primary motives for taking on an increasingly complex series of technomadic projects has been to provide context and justification for learning curves—half the fun of this stuff is diving into something you know almost nothing about and becoming expert enough to do an end run around traditional approaches and make a contribution to the field. If you're driven by personal motives like *passion*, that can happen surprisingly quickly.

You are probably already aware of an important phenomenon: the nonlinearity of learning curves and their eventual asymptotic leveling. Since you're reading this, it's safe to assume that your brain is a rather AC-coupled affair, implying that some degree of change and growth is necessary to keep you interested. If you find yourself getting bored and restless in a job (or relationship, or town...) when the slope of the learning curve approaches zero, prompting you to migrate from one situation to the next just to prevent intellectual atrophy, then you already know *exactly* what I'm talking about.

So how do we arrange our lives to keep this from happening? Easy... just take on a massive project, then redefine it and add components as required over time to keep it interesting. A nice side effect of this is that if you play your cards right, you can keep yourself surrounded by experts who know a lot more than you do about various parts of the system, even while you become the world's leading guru on the thing you're building.

The Microship project is a perfect example. I had just spent a decade focusing my life on the bike, and for all its complexity and unfinished subsystems, it had become pretty stale in my mind. Few surprises lurked in the bicycle-touring lifestyle after 17,000 miles, and even though *BEHEMOTH* was intended to be a development platform, it was hard enough to hack that I didn't find myself rushing to build new bike subsystems nearly as much as I thought I would. The truth was that I was just burned out, so when I fell in love with tiny boats I was lured wide-eyed into a whole new world of unfamiliar knowledge: navigation, hull design, oceanography, harsh-environment packaging, composite materials, and more. I started reading again, finding and hanging out with experts whose language I could barely understand, slowly building a dream evocative enough to attract the people from whom I most needed to learn. This was profoundly energizing.

I subscribed to nautical magazines, loaded up on books, and hit trade shows. I posted questions to newsgroups and forums, built up a stable of advisors, and seized every opportunity to get on the water in boats of any size and configuration—learning every time. I played experts against each other by striking up 3-way email exchanges on diverse topics like HF antenna coupling to a rotating mast, participating in discussions with people who have spent years accumulating arcane wisdom far beyond my own, interjecting questions to keep things moving in the right direction. I was careful to establish engineering contacts within sponsoring companies, learning as much as possible about the constraints, problems, and unexplored potential inherent in their product design (stuff that's hard to get from marketroids). I even leveraged my bike-era notoriety to cozy up to authors and recognized gurus in these unfamiliar new fields, sometimes embarrassing myself with stupid questions but in the process building an incredibly potent database of experts—most of whom were remarkably patient with the early phases of my simultaneous learning curves because the *dream itself* had sparked their curiosity.

Although there is some risk in issuing "forward-looking statements," as corporate types like to say, I have also found considerable educational value in opening the design process to public scrutiny. I blog[1] weekly about the ongoing development of my new ship (*Nomadness*), and am not at all shy about discussing plans... or admitting my uncertainty about how to address a problem.

Surprisingly often, this is all it takes to pull answers out of the ether. Somebody who knows way more than I do about the subject will submit a blog comment, send an email, or utter a quick tweet... basically giving me the benefit of their experience with the problem that has me stymied. Whole comment threads (and in some cases, friendships) have emerged from this, which would have been impossible had I waited until a design was a *fait accompli* before going public with the details.

---

[1] *http://nomadness.com/blog*

It can be embarrassing, of course... a major hazard of life on the bleeding edge is that one is a newbie much of the time. But I will accept the occasional public cringe if it prevents a large-scale screwup. More than once, an idea that I just *knew* was brilliant got shot down by someone who had spent a whole career developing intuition about the subject. This was painful at the time, but I realized in retrospect that I was saved from pouring resources into any more dead ends than absolutely necessary.

The counter-argument, of course, is that often said experts are inculturated in old methods, and can't resist snorting in derision at the exuberant yammerings of a noob who doesn't even know what's impossible... but who ends up successfully breaking all the rules with a crazy design that nobody took seriously. You have to listen to even the most constructive criticism with a finely tuned discriminator, and not be afraid to stay the course if you're *really sure* you're on to something. Just don't let ego get in the way, especially early in the process when you aren't yet sure how much you don't know.

All those dynamics, coupled with IP-centric nervousness about revealing potentially valuable new ideas, conspire to enforce paranoid secrecy in the early stages of a project. But if you are serious about maximizing free education, try to publish early and often... it will attract teachers of every form.

Chapter 3

Sponsorship

This is a subject that gets everybody excited. Every few months, I receive email from somebody who has surfed across my list of hundreds of sponsors, basically asking: "How do I get companies to give *me* stuff for free?"

I've probably dashed a few hopes with terse responses that convey basic realities, the most important of which is rather obvious: the only realistic sponsorship deal is a win-win. Vague plans for media coverage are usually not enough to convince people with budgets that they should send out free goodies, and some companies that make particularly alluring adventure tools receive so many proposals (on the order of dozens per week) that they have instituted formal procedures and tough filters; others have adopted a firm policy of only donating to nonprofit corporations.

Yet, there are times when it feels like Christmas every few days around here, with the UPS and Fedex guys dropping off new toys and a half-dozen new proposals always in the works. How do we do it?

3.1 Product versus Cash

First, let's get one thing straight. Asking companies for money is a *lot* harder than asking them for products, and if you plan to use the money to buy products anyway, you might as well eliminate those troublesome intermediate value states and their associated end-of-year implications. Besides, when you ask for money you end up dealing with the kinds of people who think in terms of return on investment, deliverables, tax laws, penalty clauses, and all that other weird stuff that causes the eyes of most geeks to glaze over. But when you go looking for goodies, you find yourself talking with engineers (sometimes even kindred spirits who wish *they* were doing something so cool), company principals (they may be suits, but they can make the instant decision to support you without having to sell the idea upstairs), and the aforementioned marketroids (hey, they may not always be the most creative folk, but they are usually friendly people who understand the value of good PR—and can be very helpful).

There is a counter-argument that applies in some situations: a big "title sponsor" who gives you a pile of dollars translates onto only *one* relationship to maintain. Seeking a separate sponsorship deal for every component can gobble months of schmoozing time, and you can even find yourself having to delicately balance relationships with companies who see each other as competitors. This can get a little tricky, especially where IP or pre-release products are involved.

Between the bikes and the boats, we've had well over 200 corporate sponsors, and with only two exceptions they have all provided products or services, not dollars. The two who provided money did so in a very structured way—one as a consulting contract with a defined (albeit very liberal) deliverable, the other in the form of covering the lease on a lab building. No company has ever just handed us a lump sum of cash, although we've had a few individual donors.

Some projects (like alpine "first ascents") seem to need direct financial support to pay for people or logistics, but from my perspective that appears to be a much more stressful relationship, both in terms of the difficulty of finding major sponsors and providing a return that fulfills contract terms. This typically involves large-scale logo display, scheduled appearances, or serious product placement. The problem with all that is that it can really define your media image... and be difficult to change without stepping on toes if you suddenly find another brand you prefer (sports stars do this dance all the time). Best to sidestep the whole issue if you can.

In the early days of my bicycle travels, I was offered a significant monthly stipend—more than I had ever made in my life, although that's not as dramatic as it sounds—to pedal an epic South American journey while promoting a particular brand of tobacco. As a militant anti-smoker, it was not difficult to politely decline, though I was terribly broke at the time and did timidly ask if they could launder it through their beer subsidiary (they were not amused). That was an easy decision, and it's a good thing: other than the obvious hypocrisy of promoting cigarettes on a bicycle tour, the one-shot hype experience would almost inevitably have prevented my nomadness from becoming self-supporting over the long haul.

More interestingly, however, it would also not have been fundamentally different from visibly identifying with, say, Apple or Motorola, companies that make things I actually *do* use. How would the media ever take me seriously, especially if I was talking about computers and communication tools? A big logo on the side of my machine would taint every statement with obvious bias.

This is the other key reason to try to avoid such relationships: as we shall see later, the media is a critical part of a three-way symbiosis with the project and the sponsors, and it wouldn't do to look too much like an advertisement. Obviously there is some product promotion going on, but fer chrissake, let's at least try to be *subtle* about it!

If we're not going to do logo decoupage all over the hull, what *do* we offer sponsors in return for their generosity?

3.2 Win-Win Relationships

If we think creatively, there are lots of things we can do to help a company that has just donated an essential component to a gonzo engineering project. Frequently, the sponsors are not high-profile purveyors of whizzy consumer goods, but are small, specialized companies that target their wares to industry. In these cases, a little publicity goes a long way, as they're generally not used to this sort of thing and tend to be more receptive than outfits that get a stack of proposals every day. This is a key point, so I'll say it again: don't beat your head against the gates of Microsoft, offering to trade your immortal soul by clicking through the EULA on a free piece of remote-admin spyware with auxiliary productivity features, when some small outfit making valves in Mississippi will fall all over itself for the opportunity to ship you parts in exchange for a little free ink in the trade press.

Media Exposure

"Ink," of course, is the first thing people think of when considering what to do for a sponsor: name mentions during interviews, product-application feature stories in industry trade journals, photos that show their gadgets being used in an unusual way, links on the project website. We pursue every opportunity to do all these things, and some companies with particularly visible or mission-critical components have received piles of media photocopies over the years—name mentions highlighted in yellow and a nice thank-you letter paper-clipped to the top.

But sometimes this doesn't work very well. Some products are in the class of development tools or are so deeply buried inside subassemblies that they couldn't be photographed effectively if we tried, and mentioning them in general media interviews would just confuse reporters ("This is the Microship, which uses stacks of DuPont Hytrel crane bumpers from Miner Elastomer as landing-gear shock absorbers.") Other than slipping plugs into obscure publications like I just did, how can I make sure that all sponsors' investments are worthwhile—increasing the odds that they will still be receptive if I need to hit them up for more goodies next year?

The first (and easiest) solution is to maintain enough of a public project narrative that everything is automatically included. Feature-length magazine articles are terribly restrictive in length and slant, and a recitation of vendors is boring to readers. But we have a blog, busy websites, and over 5,000 subscribers to an occasional emailed *Nomadness* posting, and as far as I know every sponsor has been mentioned at least once along with a paragraph or two about the product and a link. These are archived on our site, and are thus Googlable; there is also a sponsor page listing them all[2], and for many years, entry to the Microship website triggered a Perl script that served up a

[2] *http://microship.com/sponsor*

random logo. The *BEHEMOTH* bicycle even had a sponsor-logo slideshow that ran on the console Macintosh during media events.

Still, that's not much exposure, although it is appreciated.

Ideally, I try to do a series of magazine articles in a variety of markets, highlighting each part of the system. This generates a bit of cash flow, while providing lots of opportunities to mention companies within the context of a trade journal targeted to their industry. It's harder than it sounds, however, as each one has to be pitched and sold to the editors, not to mention the problem of including enough of a general introduction to bring readers up to speed. In practice, this is not an effective way of making sure that every sponsor gets useful return for their donations, as only about half seem to lend themselves to this approach: cutting-edge newsworthy technologies, or highly visible components that show up clearly in photographs. Let's see what *else* we can do to make sponsors feel good about helping out with the project.

Enhancing Corporate Culture

Sometimes the decision to make "in-kind product donations" is so casual that companies don't really expect a lot of flashy media exposure in return... just coverage in their own publications. I have done dozens of interviews or provided photos for various sponsors' *house organs* (internal newsletters or magazines). A perfectly reasonable motive for supporting something exciting like this is to get the employees pumped about cool uses for their technology, and this actually can take a variety of forms:

I have camped for months in corporate facilities, building systems and deriving direct support from the employee population. (In such situations, I became a sort of high-tech court jester, adding a buzz of excitement to the routine 40-hour-a-week *blahs.*) I have conducted brown-bag luncheons for companies that donated equipment, giving energetic speeches that cover the entire project while emphasizing how their products were indispensable components and relating amusing anecdotes about integration or comments from the public. I've shown up at company picnics, sent regular updates to mailing lists or forums that repost to internal mailing lists, and become friends with the engineers who created the technology that I use. In every case, the net gain back to the sponsor is clear: their employee population gets to see their products appreciated outside the normal markets, part of something exciting and fun. Team spirit is powerful stuff, and has soft-dollar barter value.

Marketing Participation

Another angle is to participate directly in product marketing, not by independently getting media coverage but by offering your image for use in advertising. Personally, I tend not to do this too much as I don't want to become strongly associated with any single company, but those fears are probably unfounded—the few times I have done it worked out well.

One amusing variation on this that works particularly well with my fancy gadgetry is to appear as a guest in the sponsor's booth at a trade show. I've done this a lot, and it's good for all concerned: the company gets a kick-ass booth draw, the attendees get something way out of the ordinary with a bit of celebrity panache, and I get to hawk books and make contacts with other potential sponsors.

At a late-80s COMDEX in Las Vegas, I was exhibiting with Chips & Technologies. It was Day 4, and I was in serious burnout—the problem with these things is that person *n* arrives and asks a level-1 question, which you politely answer. Then person *n+1* arrives a few minutes later and jumps into the conversation that's just getting interesting... "Excuse me, what are you doing here? What is this thing?" After a few days of getting reset to zero every few minutes, you want to start punching people in the nose. I was in this approximate mood when a fellow arrived in the booth, looked at my bike, and exclaimed, "Wow, this is the coolest thing here!"

"Thanks," I said, handing him a flyer and sizing him up for a book sale.

"Yeah, but not for the reasons you think."

Oh crap. Another nutcase. I was steeling myself for the brain dump, religious testimony, or sales pitch when he bent down, dragged his finger along the frame, held it up to the light, wrinkled his nose, and said something that has stuck with me for the past decade:

"Look at this, it's filthy! It's the only thing at this show that I know is *real*."

Since then, I've never underestimated the value of putting a bizarre contraption in a sponsor's trade-show booth... it can mean a lot more than just another desperate theatrical attempt to get jaded attendees to stop and get their badges scanned for the mailing list. As long as the project genuinely uses the company's products and is not just hired to be a gratuitous booth draw in the same class as bikinis and magicians, it can be hugely effective.

Field Testing

But hey, wait a minute. We're *geeks*. What's with all this marketing garf? Can't we return value to sponsors by doing something that's actually clever?

Well, I hope so. In the case of the Microship, we already have a whole infrastructure for data collection and the archiving of an arbitrary number of channels; this can be useful to some of the companies that have contributed products. Certainly it's not much trouble to record temperature cycling, humidity, and other environmental conditions—even PSD plots of the worst-case shock and vibration events. The big boys already put their stuff through its paces by subjecting samples to such abuses under much more controlled conditions, of course, but I've seen a few product specs that were compiled optimistically from the published data sheets of their components. While wandering around outdoors for a few years hardly qualifies as a proper "accelerated life-test" program, it can certainly yield useful data... especially if the extremes happen to correlate with performance anomalies. We report failure modes, experience with adhesives or mounting problems, and anything else that may help forestall future problems.

Sharing the Hacks

There's another way geeks can do something for sponsors without posting logo GIFs or grimacing into cameras (though I actually kind of enjoy that sort of thing, in moderation). Why not return all the hacks, interfaces, and driver code that you had to conjure in order to put a particular widget to use—in the process, rewarding your sponsor with a sort of off-site skunkworks that is *unconstrained by internal politics?*

Sometimes this is irrelevant, but there have been times in these projects when we've been able to give our sponsors significant intellectual property—and since I'm not trying to market this stuff myself, it's not like there's any harm in sharing the results. Quite a few of the devices integrated into the ships have been removed from original packaging, given more efficient power supplies, associated with software objects and display widgets, and hacked to add interface hooks or enrich external monitoring. All this is fed freely back to the manufacturers, along with related code. Doing this has the satisfying flavor of closing the loop, and significantly enhances a sense of co-operation with the company—the result can be a very effective partnership in which you not only get ongoing gadget upgrades, but the expertise to make them dance.

Obviously, it's a bit tricky to generalize from computerized bicycles and micro-trimarans to all possible gonzo engineering projects, and some of this is irrelevant to, say, a software undertaking or epic data-collection exercise. But even tools built entirely of bits can be fair game for sponsor relationships, and hopefully we have shown how this can turn into a win-win. The only commentary I would add at this point is this: companies are made of *humans*, somebody there is going to remember that they made a donation, and employees want to look good to management. I can't emphasize enough the importance of keeping track of your supporting companies and contacts, making sure that you give them *something*, at least an occasional update, in return. It's not only likely to help you further down the road when the original version gets superseded by a new model, but it will make things easier for the next person who approaches them for help. More than once, I have been told: "Sorry, we tried sponsoring a couple of projects a while back, but it never paid off. Our policy now is to just offer a discount."

Chapter 4

The Media Dance

I don't think my projects would have survived without all the media coverage over the years. It has opened more doors than I probably even realize, led directly to writing and speaking gigs, smoothed the way to countless sponsorships, sparked friendly waves and warm hospitality offers, and probably even saved my life on narrow roads ("Goddamn, slow down a second, Bubba, there's that computer dude I heard about on the TV!"). Though one could certainly make cynical observations about all this, it is true that in our media-centric culture a little fame goes a long way—although it does not, alas, automatically imply fortune. Trust me on this.

Like sponsorship, media presence is an essential component of a gonzo engineering project; it has a way of making everything else easier. But there is an art to it... don't assume that just because you're doing something original, reporters will flock to your lab and write gushing stories. Let's explore this a bit.

4.1 The Digestive Tract of a Horse

One of the most important things to remember is that you have to somehow manage to be newsworthy without drifting too far into the domain of self-promotion: there is nothing more tedious than someone who is always tootling his own horn and viewing the world through the filter of his own personal obsession. Media professionals have a finely tuned radar for this sort of thing—as well as blatant product promotion. They know when they're being manipulated, and it can backfire badly since *they* want to be in charge of the story and do not welcome attempts on the part of the subject to run the show.

But with a little care, you can use this to your advantage... just make sure that you're providing a clear and consistent image rather than amorphous raw material that can be digested into a different story entirely. This happened to us one year at the Hackers Conference; a crew from *CBS News* showed up, took a few hours of footage, and produced a completely unrelated piece on computer crime illustrated with B-roll of some of the industry's leading luminaries in the woods above Silicon Valley. The media were never invited back, of course, but damage bordering on libel had already been done.

Admittedly, a whole conference is a bit hard to control, and the wide range of colorful personalities gave them enough material to support just about any fiction; they could have done a piece on the favorite hobbies of extraterrestrials had they so desired, armed as they were with both a serious shortage of knowledge and extensive video of übergeeks at play.

Fortunately, as the brains behind a gonzo engineering project, you have the opportunity to keep a much tighter rein on the media beast... it can even be your friend.

I was lucky... my first lesson along these lines was delivered in a helpful and non-destructive way. I was doing an interview in 1984 with *CBS Morning News*, my first national television appearance. Unlike their evening-news counterparts, this team was professional, their production standards were high, and we got along beautifully. But still I was in for a shock: with the cameras rolling, the interviewer asked how the computer networks figured into my bicycle-touring life. I launched into a long and precise explanation, filled with rhapsodic asides and pithy anecdotes.

"Cut!" cried the producer. "I can't use any of that. You're referring forward to things you haven't said yet and back to things you said five minutes ago—how am I supposed to edit? *I want 20-second speech bites, ending on a downward inflection with your mouth closed."* This was delivered with a grin, so I knew he was yanking my chain a bit, but there was truth in his words. As painful as it is, you have to be able to package even the most abstract concepts in clear, memorable packets... pausing in between to suggest natural edit points. And you need to do it in a way that still sounds like natural speech, not like a rehearsed stand-up delivered by a news reporter. It's a fine line, and takes practice.

The metaphor to remember is that *the media is a big horse:* if you stuff straw into one end, something remarkably unlike straw comes out the other. But if you insert bricks, you get bricks out the other end. You need to develop concise comments and colorful sayings that collectively define your work, then casually speak in those during interviews the way a Zen master speaks in koans.

We can extend the media-as-horse metaphor a bit further: it is always hungry, and this works to your advantage. Every minute of air time, every column-inch in the newspaper, and every page of editorial space in magazines has to be filled with content... day after day, month after month. Coupled with fierce competition at every level, this translates into a lot of media channels just aching to tell your story *if* it's sufficiently interesting. The trick is giving them a heads-up without making it obvious that you *want* media coverage; as in the dating scene, there's nothing like an air of desperation to send people scurrying in the opposite direction. I've had hundreds of articles and interviews, but have never called a station or publication to offer an interview—not once.

So what's the trick?

Well, think about it from the reporter's perspective. Deadline is approaching, and there's a hole to fill. What's hot, new, visual, funny, poignant, disturbing, or likely to yield a Pulitzer Prize? You develop a keen eye for fresh material: trolling the web, listening for scuttlebut, and *scanning other publications in non-overlapping markets.* Do you see the implications of this? Sometimes all it takes is one or two stories to trigger a whole series—I've had otherwise unremarkable local news items explode within days to a spot on CNN, calls from feature-writers for glossy trade magazines, requests for photos and interviews from international business publications, and an hour on the Donahue show. Starting the process can be as simple as having someone else call in a tip, or visibly doing what you do and letting them notice you. Occasionally, the symbiosis with sponsors can come into play here, as companies who actually sell things are less shy about initiating contact with the press and may welcome the news hook that you represent.

If you *do* need to draw attention to yourself, do it in the form of a low-key press release with an absolute minimum of hype and exaggeration. They've seen it all, believe me, and anything that looks blatantly self-promotional or less than professional goes straight to the recycling bin.

Basically, the only trick is getting started... and then keeping the story interesting. Let's look at this a bit more closely.

4.2 The Project Moniker

Much of the art of being interesting to the media is simply having an image that can be grasped and communicated in a short news segment. Think about a spot on the evening news, or a 300-word "tight and bright" piece in the inside pages of *USA Today*: if it takes an hour to explain to a fellow geek what you're doing, it might be impossible to get it across effectively to a bright-eyed reporter whose last assignment involved converting a consumer product press release into a photo/caption puff piece. In other words, those sound bites we talked about above aren't just for use during the artificial reality of being on camera; you have to be able to convey the essence of your project in few words with startling clarity, tweak it in various compelling ways to appeal to different markets, then keep it dynamic as the years pass so it doesn't fade into the static.

The entertainment industry has a name for this: *the elevator pitch.* If you can't get the story idea across in the time it takes to ride the elevator, then it is too complicated.

An important component of this, as trivial as it may sound, is the project name. You need a memorable handle, and one unique enough that even the most watered-down article or happy-news blurb will leave people with a way to find out more by typing something into Google. Otherwise, you lose one of the most important side-effects of media coverage—the ability to attract volunteers, sponsors, and growing image recognition. Consider these two hypothetical spots about the same fictional project, delivered in the voice of an evening-news anchor:

> "And finally this evening, there's a man in the U district who's been working for three years on a way to help people interact with computers using a whole variety of gestures, not just typing on a keyboard and manipulating a mouse. He uses cameras and sensors to keep track of the movements of a user's hands, allowing the creation of a customized visual language to increase the efficiency of man-machine communication."

> "And finally this evening, meet the *Gesturizer*. This remarkable device was invented by a man in the U district, and it lets you interact with your computer the way the deaf communicate with sign language. He says it takes the average user only two days to double the efficiency of computer use."

By coming up with a catchy name (even a silly one like "Gesturizer," which does not exist at this writing) our subject has created a clear label that insures every news story will unerringly point to his website, or at least be easily searchable. When you hear another story about this project six months later, it will sound familiar; with any luck, you've even bookmarked gesturizer.com in your browser or signed up for his occasional email updates. And something with a quirky and memorable name certainly makes for a more lively news story than the abstract explanation in the first example.

Notice also, in the second snippet, that there is something like a sound bite nestled in there. It's a fair bet that the reporter scribbled "interact w/ comp like deaf w/ sign language" during the interview, knowing that a visual image would make a good story lead. There is even a startling statistic to further capture the imagination: wow, only a couple of days' practice to double my efficiency? Where do I sign?

By maintaining steady media coverage for the Gesturizer over a few years, our fictional developer enjoys the luxury of growing image recognition, thus creating a context for new stories and updates:

> "You may recall the amazing *Gesturizer* that we reported on last Spring. Well, the inventor now reports that his system can actually recognize sign language from across the room, no longer requiring a computer user to remain tethered to a keyboard and mouse—or even a desk! Our science reporter, Bob Archibald, has the story..."

Without a memorable name and a tasty sound bite, this follow-up story would have had a much less compelling introduction.

This sort of thing may be anathema to geeks who want to remain focused on system development, not activities that seem better suited to a marketing department. Clever names, websites, image-management, media spin, sound bites... what does all this have to do with gonzo engineering? Believe me, it pays off, even if you haven't the slightest intention of ever selling anything—you can't run a massive project in a vacuum, and the media is your best ticket to meme-propagation.

4.3 The Art of the Demo

If you thought that was bad, now I'm really going to make you gag. When you're dealing with mainstream media instead of highly-targeted journals in your native field, you have to allow for the possibility that they might not have the slightest clue what you're talking about. I've actually been on the other side of this; the first time I covered an Artificial Intelligence conference for *Byte*, it took a while for me to see natural language systems as anything more than fancy incarnations of Eliza. (That particular subculture had a real PR problem—most of their publishing activities were incestuously contained within the AI community itself, and the general media was getting a distorted, hype-filled view of what was going on... creating inflated expectations that were destined to lead to disappointment.)

The problem here is that when the topic is a sufficiently interesting system, there is usually a rich shared context with the audience that you can normally just assume. This is why publishing technical articles is easy; you already know that your geek readers understand the prior art, and will thus *get* your clever hack or new paradigm if you explain it in that context. But now sit a writer for the *Podunk Weekly News* down at the console and perform your usual demo. Blank stares? Tentative mouse clicks on irrelevant objects? If you're hearing questions like, "So, um, is this thing connected up to the Internet? Does it get TV? Who would actually use it?" then you need to create another class of demo.

This was easy and fun with *Winnebiko* and *BEHEMOTH*—showing the act of writing-while-riding was not particularly dramatic, though the more clever reporters identified with it immediately. (I am, after all, in their business... yet not enslaved to a desk.) But I needed a few cute things that would play well on camera while conveying the essence of the system—the operational equivalent of a sound bite. I coded up some simple scripts that could be remotely invoked with keypad commands via a handheld ham-radio transceiver, and used those to orchestrate scenarios that worked particularly well with the TV news.

In North Carolina one afternoon, a reporter stepped in front of my bike and began his summarizing "stand-up." As his shadow fell across the solar array, I surreptitiously sent a Touch-Tone command and the bike's synthesizer sternly spoke: "Excuse me. You are blocking my sunlight."

It was priceless... still on camera, he quickly stepped away, saying, "Oh, I'm sorry."

I had a canned collection of remotely triggerable speech strings, and in almost every interview they came in handy. There were also some large-font files (themselves serving as "text bites") that I would pop up and edit when demonstrating the handlebar keyboard, eliminating the *quick brown fox* or worse, the *fsdfsfdsfs* syndrome. A couple of favorite CDs were always at hand to show off the music system; I became adept at popping the top off the Qualcomm OmniTRACS satellite terminal and demonstrating azimuthal tracking upon loss and reacquisition of signal; I would hand the reporter my sweaty helmet and let them see their own contact information in the heads-up display. Coupled with ongoing patter about the uses of the underlying technology while on a bicycle tour and its implications for future widespread nomadic connectivity, this sort of thing added visual appeal and contributed to my ability to drive the story with consistent content, rather than just answer the usual questions (What did that cost to build? Has anyone tried to steal it? Is it hard to pedal up hill? Have you had any accidents? Where do you sleep? What do you do in the rain?)

4.4 Leveraging the Media Portfolio

Over time, with a little care and luck, a sufficiently interesting project will begin to accumulate clips—newspaper articles, tear sheets from magazines, and a collection of video dubs. This stuff is gold, and can serve you well for years if you use it wisely.

First, keep good records, both in the form of a mailable document and a list on your website[3]. Media coverage is one of those things, like credit, that exhibits positive feedback: the more you have, the easier it is to get more. Any doubts a reporter may have about your credibility will evaporate in the face of a good list of previous coverage, although there is a limiting factor in that some publications are sensitive about being "scooped." When I first passed through Seattle on the bike in 1986, I interviewed with both of the daily papers... and the *Post-Intelligencer* dropped the story when they found out I had already been on the front page of the *Times*. (That was annoying, but not as bad as the New Orleans *Times-Picayune* that killed the story because I didn't have a "handicap" that would make a cross-country bicycle trip newsworthy!) But in the larger world of magazines, this has rarely been a problem; there is always a different angle available, and the value of a story becomes amplified by the celebrity effect.

---

[3] *http://microship.com/press/media*

Another reason to keep these lists, of course, is to facilitate sponsorship proposals. Even if someone hasn't already read about you somewhere, they are sure to be impressed by a long list of publication credits backed up with a few carefully selected photocopies. The collection becomes immensely valuable and will be important to you for most of your life, so put those articles in protective sleeves and keep them in binders! The best ones should even have backups.

Note how this keeps looking like a 3-way symbiosis: sponsors get media coverage, you get goodies, the media gets stories about interesting technology. All three components have to be carefully balanced, ultimately driven by the energy of the project itself. If you plan it well and keep it evolving in interesting directions, it can even become self-sustaining.

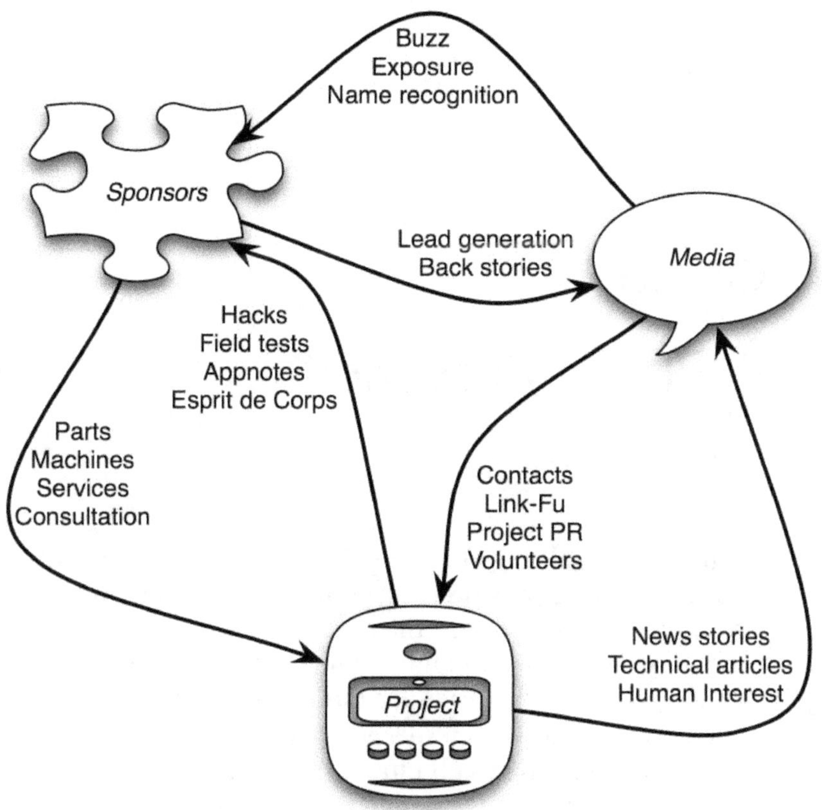

Three-way symbiosis of project, sponsors, and media. Everybody wins as long as the process keeps moving forward, and it is up to you to drive that. New toys are inspiring; build something amazing with them, and the press will tell your story. If you craft interviews to honestly underscore the role of your sponsor(s) without blatant flag-flying, then you can keep the outer loop going indefinitely. The hard part is the leap of faith on all sides necessary to get it started.

Inner, direct connections are just as important, and the most productive relationships contain all the elements shown here.

Chapter 5

A Public Presence

We have been talking about a public presence already, of course, but media coverage is a unique phenomenon: although you can increase the odds of accuracy and a positive spin with a good name and some clever sound bites, the published stories are out of your control. If some reporter thinks you're a colorful nutcase and gives the story a snide "takes all kinds!" slant, there's not really much you can do about it... and it certainly isn't going to help build a positive project image. (Don't believe the old nonsense about any publicity being good as long as they spell your name right!)

There is another aspect to the public face of the project, however, and over this you have complete control. Thanks to the Internet and common design tools, anybody can be a publisher these days... if you were born after 1980 or so, you may find it hard to believe that there was a time when every geek didn't have a website with content-management software, mailing list, blog, and a portfolio of domain names reserving his or her pet neologisms. In the Olden Days, we actually *typed* things directly onto paper, character-by-character (sometimes even paying professional "typesetters" if something needed large fonts or right-justification), then produced expensive *print runs* of hardcopy publications. Distribution, in the pre-Net era, involved affixing colorful gummed receipts for the transportation charges, then handing over these "stamped" hardcopies, sorted by zip code, to uniformed workers who would transport them to their destinations in a sort of physical analogue of packet-switching. <creak> Am I dating myself?[4]

It almost goes without saying that any gonzo engineering project should have a website, updated frequently with news, photos, downloadables, YouTube clips, bloggage, white papers, personnel sketches, public appearances, media coverage, sponsor lists with links, and related resources. I don't really need to tell you how or why to do this. Just do it.

Beyond that, I highly recommend setting up a few email lists. Some members of "your public" will in fact be proactive enough to check your website regularly (or use RSS) to see what's new, but there's a lot of cool stuff out there competing for mind share and, at least in my own overloaded life, I find that I rarely have time to "web surf" anymore. If something is not immediately relevant or referred by a trusted source, I am not likely to stumble across it in the act of poking around... the thousands of bookmarks accumulated since the "Mosaic" era do me little good. If I have a question, I Google my way to the answer; I read a few regular news-aggregators and blogs of current interest; I scan the dozen or so forums that manage to hold my attention. This is probably typical.

But a mailing list is the original "push" technology. Once someone expresses interest in your project, *you* get to

---

[4] No, but I am sitting by the phone waiting for me to call.

drive the update process. As long as your content is interesting and you don't start spamming or posting too frequently, that person will probably stay on your list throughout the life of the project. Admin and bouncie-processing is a pain, but it's worth it; my *nomadness* list, now over 15 years old, is one of my most important assets. Among the thousands of names are some amazing people: almost without fail, I can mention a conundrum in one of my updates and someone will provide the answer within hours. The list includes sponsors, writers, CEOs, friends, volunteers, engineers, marine architects, sailors, advisors, potential hospitality sites, the curious, the skeptical, and a few non-English speakers who probably thought they were signing up for information about Microchip Technologies PIC development tools and wonder why they occasionally get rambling email about some *barco ridículo*.

This list is transmit-only, meaning that I am the only person who can post. With thousands of names, the usual discussion and noise could easily get out of hand ("how do I get the *%$! off this list?" followed by a wave of flames), so I restrict this "channel" to pure publishing. But there are other needs as well...

I maintain active discussion lists for technomads in general, participants in the planned flotilla of boatlets, and developers working on the project. Multiple dev- lists come and go over time to segment the latter into more focused subgroups. (I should point out to the same youngsters who had to be told about paper mail that there was a time when, for practical reasons, volunteers for a project had to be in the same town—not scattered across the planet and unlikely to ever meet face-to-face.)

In addition to websites and mailing lists, it is advisable to have at least one reasonably well-produced print publication that describes the project. Over the years, I've tried to put out quarterly journals, but that's too much like work (though I once told impatient readers that I actually meant "every quarter-decade"); more successful has been a small series of technical monographs whose flagship is a 110-page self-published book[5] that gives a light overview of the whole series of bikes and boats. It's not highly technical, but it does generate a few nickels while doubling as a presentation piece for those who wander by the lab to help. It's even a useful resource for reporters, letting them dig around for quotes at their leisure instead of recycling my comfy old sound bites again and again.

For a while, I ambitiously planned to expand the monograph series to a whole product line covering all aspects of the Microship project, but doing so is time-consuming and expensive (and it's way too easy to end up with either a pile of obsolete inventory or a lot of wasted time designing "publishing on demand" documents that were never, um, demanded). The alternative is easy, fun, and has a "cost of goods sold" that is arbitrarily close to zero: a library of PDF documents tucked into a shopping cart that takes PayPal. These can be sold or given away, further spreading the memes of the project and maybe even generating another nickel or two.

Coupled with appearances in more traditional media, these various activities generate a rich flow of outgoing information that gives a project the appearance of healthy activity... an essential image to maintain. Let's look at one last component in the complex blend that is necessary to propel a dream past inertial effects and potential barriers, maximizing its chances of reaching escape velocity.

---

[5] *From BEHEMOTH to Microship* (ISBN 1-929470-00-2)

Chapter 6

Building the Team

If you're at all normal (which I cannot assume, but hey, we're talking *statistics* here), you can only work in isolation for a limited time before reaching burnout—whether from loneliness, overload, or a lack of constructive feedback. It's analogous to music: I've played the flute all my adult life and have a familiar set of ruts that I fall into when jamming by myself. But shove me in front of a microphone with other musicians, and after a few fumbling minutes of feeling like an incompetent klutz, I find myself playing fresh material. I recently started the piano, just to keep my lazy old wetware on its toes in new territory.

It's like this with the design process, and not just for the practical reason of distributing workload. Everybody brings ideas and experience to the party, and when you're in the envelope-pushing business you need all the inspiration you can get. Even for those rare folks who indeed know it all, proceeding indefinitely without intelligent company is just plain difficult. The trick is recognizing and attracting the kinds of people who have a lot to offer, and it's no accident that those are the very folks who are already overloaded and unlikely to list "volunteerism" as a hobby. Just as with sponsor relationships, this has to be a win-win, and that can take a lot of unconventional forms. (It can also be based on money or barter; we'll get to that in a moment.)

6.1 Attracting a Volunteer Community

What can a gonzo engineering project offer potential co-conspirators, assuming there's *not* a big pot of cash available for hiring first-class talent? In my projects over the years, there have been a few "perks," some accidental, that have kept me surrounded with interesting people:

For the Glory!

Perhaps the most obvious benefit that a volunteer crew can derive from pouring creative energy into a project is the resumé-enhancing halo effect of association with something exciting. It is thus exceedingly important that you don't greedily hold on to all the credit, but instead share the spotlight with those who have made contributions. In my own case, I actually get a lot of pleasure from this: when I kick back with a beer and contemplate the bike or boat, I see nearly two decades of intelligent friends, mad marathons of midnight machining, student teams working with rising panic as the end of a quarter nears, and pivotal moments of insight contributed by people who haven't been blinded (as I often am) by staring at the same problem for months. In articles, interviews, and speeches, I give credit where it's due.

Of course, offering this as a sort of barter is a bit iffy; if a project is just getting started, promises of fame come off a bit like overly ambitious hand-waving, like a startup promising founders' pool to someone who helps cobble together the “About Us” HTML. It takes a bit of time for this to become as attractive as it sounds.

The Bottom Line is Fun

All that seductive fame ‘n glory aside, I have found over the years that the best motivation happens to be the most pure: fun. This is, after all, why most techies do what they do, and one of the best compliments I ever received on the project was from a fellow who dropped by for an afternoon of brainstorming: *“Thanks for reminding me why I became an engineer.”*

I am convinced that the best way to build a team is to simply do something so cool, so exciting, that people jump at the chance to participate. This is the kind of energy that gets creative juices pumping, and feeds the project... I dunno about you, but after a few years of pushing toward the same goal I tend to get a bit jaded. There’s nothing like the arrival of someone who’s turned on by the Microship fantasy to get fresh ideas percolating and knock a few jobs off the list (especially TBDWLs[6]).

Geek’s Vacations

It occurred to me a few years ago that I could carry this amusement theme to the next level by formalizing a “geek’s vacation program.” It’s simple: we invite clever folks to come stay with us for a week or so, taking a break from their normal jobs to do design or fabrication work for the pure pleasure of it. We provide food, beverages, a frolicsome atmosphere, and a few outings of playing tourist here in the Pacific Northwest. In exchange, we get the

---

[6] *To Be Dealt With Later*

focused intellectual energy of some truly amazing people, and always manage to learn something.

Social Engineering

It didn't occur to me in the Early Days, but another one of the most alluring features of these projects, from the perspective of our volunteers, has been purely social. Once it reached critical mass, with enough participants to make overlap likely, a new component entered the picture: folks would come by the lab to help, then meet each other and form new friendships. This has led to businesses, lasting relationships, and uncountable pizza nights... I'm not the only one who tires of working alone, it seems. If you make your Gonzo Engineering project entertaining enough to lure clever people, you'll find that it might take on a life of its own. That's where the real magic begins.

A skunkworks-like microculture developed around my projects, attracting programmers and machinists, hams and cyclists, human-powered vehicle gurus and chip designers. I recall one night a few weeks before the long-awaited launch of *BEHEMOTH* back in 1990... the stereo jamming, the windowless lab at Sun Microsystems a sea of fluorescent-lit clutter, the SPARC beeping every few minutes with incoming mail as I worked to nail down the logistics of the tour. Michael Perry was writing FORTH code to drive the audio crossbar, Steve Sergeant was chasing a noise problem, Steve desJardins was working on the landing gear 4-bar linkage, and Zonker Harris was squinting into a dense matrix of Lemo connector pins with a soldering iron in his hand. Over at the Rockwell milling machine named *Cecil* (Cecil be da Mill), David Berkstresser was standing in a sea of aluminum chips, conjuring a piece of structural artwork for the bike's trailer hitch. I heard the mill spin down, then David hollered, "Hey! Does it ever make you feel funny that so many people are working so hard to get you out of town?"

Education

There is yet one more "intangible" motivation before we get to some of the more traditional methods of luring people into your lair. If you're doing something that's pushing envelopes and is exciting to media and sponsors, it's a fair bet that it represents a very real learning opportunity. We've had people offer to help with "anything at all, even sanding," just for the chance to hang around the lab and absorb techniques or arcane knowledge. The irony here is that this is one of my main motives as well, and is perhaps the source of my greatest pleasure in working with volunteers. In the process of grappling with interesting problems, everybody brings something to the party and the educational opportunities are endless.

Sharing the Wealth

And then there's the stuff that *really* makes a geek's eyes light up: new toys. In the midst of all your rhapsodizing to a potential volunteer about education, networking, and fun, it never hurts to mention that there may be excess goodies to share. I've had leftover antennas, computers, radios, nautical components, and more—much of which has found a new home with some of our star volunteers as an extra "thank you" for all the help. There's an ethical component here as well: quite a bit of this is sponsored, and it feels a bit weird to take donated stuff and hawk it on eBay (I've done it a few times, but only with explicit approval from the sponsor or in cases where the company no longer exists). Passing along a few extras to people who have participated in the project feels reasonably consistent with the spirit of the gifts, and certainly generates a lot of smiles. So do team T-shirts.

6.2 Paying for It

Despite all the motivations folks might have for hanging around and helping you for free, however, we still have to accept the annoying truth that people have other things to do with their time. This translates into having to find more traditional ways to get help... and if you're anything like me, your dream project probably doesn't include a budget for full-time employees. For a while during the dotcom boom, I was able to make enough from corporate speaking gigs to keep a composites guru on site, but even that was a non-traditional deal that included room and marginal board (in an old camper) as part of the consultancy package. Actual employees are another matter entirely, opening up regulatory cans o' worms, workman's comp insurance, and the expectation of timely paychecks.

But there are alternatives.

First, if the associated business model we discussed earlier is generating cash flow, you can "share the wealth" by giving someone a piece of the action to take an active role. Given that the project itself is coupled synergistically with the nickel-generator, this may translate into on-site help that can keep you moving through the creative doldrums.

Second, if you have a tight relationship with a sponsor and are in the process of conjuring something that the company might be interested in leveraging, then you might be able to bargain for time from one of *their* employees. This is great, as you don't have to play business-owner any more than necessary, and if the project doesn't reach escape velocity you haven't thrown a wrench into anyone else's career.

I would be careful with bartering a vague "partnership" unless the project has a very real deliverable and corresponding business model. This is usually not the case with the kinds of passionate pursuits we are considering, and things can change a lot between concept and execution. If someone has been faithfully putting in the hours with the promise of residuals from future patent or book royalties, and instead gets to watch you getting Slashdotted and BoingBoing'd all by yourself, you could have anything from a lost friendship to a lawsuit on your hands (depending on how close your understanding was to a "legal contract).

Seriously, be careful with this. Don't get into that kind of debt. Or any kind, if you can help it.

6.3 Tapping Academia

There is one other way. Lemme tell you a tale of Microship development...

In the Spring of '93, I arrived in San Diego to speak at Qualcomm, one of my favorite sponsors—the creators of the bike's satellite communication system and gurus of CDMA and other DSP-flavored wizardry, a company managed by engineers instead of MBAs. A friend there pointed me to nearby Scripps Institution of Oceanography, where I might have a chance to pick up a few tricks from guys who know how to keep electronics working at sea (a non-trivial problem, with delicate high-impedance circuitry surrounded by conductive *aqua regia*). Armed with an introduction, I wandered over and did a casual bike show 'n tell for a Friday afternoon Scripps cookout.

As I was standing by the keg nursing a pint, my bike parked a few feet away and the surf pounding just beyond, a classic Hollywood stereotype of the brilliant young PhD walked over and said, “Fascinating. Have you gotten any results yet?”

“Well, I’ve fallen in love a few times,” I answered.

He looked at me blankly for a moment and then cracked up. “No, seriously...”

This was my introduction to academia. I hung around a few days, seeking space at the Scripps facilities, eventually managing to arrange a secure place to park the diesel Mothership and stash my gear in a run-down storage facility in a desolate nearby canyon... but with no net connection, phone, or actual workspace. It was an excellent toehold, though, and the chain of contacts kept growing as I explored the couch circuit. A few days later I met with a professor in the Electronic & Control Engineering department at UCSD (University of California).

Almost immediately a potential win-win emerged. Although it wasn’t widely viewed as a problem among the academics, there was a serious shortage of hands-on engineering opportunity for students: they could actually make it through four years and get an undergraduate degree without ever picking up a soldering iron. I found this shocking, and had a solution to offer: “Hey, give me some lab space and I’ll teach a senior projects class.”

I stumbled into university culture with naiveté and an utter disregard for hierarchies—to me, there was no essential difference between deans, chancellors, TAs, and professors. Only much later did I understand that there is a pervasive caste system, which explained my tendency to ruffle feathers. Other than holding on to lab space long enough to get the Microship built, however, I had no pretensions of a career, publishing in refereed journals, or chasing the holy grail of tenure.

Nor did I anticipate the complex politics of space wars. My first lab was an abandoned bookstore... but the day after I installed a monitored security system and set up a working facility, a construction crew hacked straight through the drywall, knocked down shelving, looked around in annoyance at my stuff, and asked me what the hell I was doing there. "We're tearing out this end of the building and you gotta get this crap outta here NOW, before our contract penalty clause kicks in and costs us a thousand bucks a day. Which is what it's gonna cost *you.*" But this disaster was tantamount to a back door from the institution-hacking perspective: now that finding space for me could be defined as an *emergency*, I landed within a few hours in the Microwave Lab, vacant for the quarter and perfect for my needs.

I moved out two years later.

Occasionally *real* professors wondered aloud what some rube without a degree was doing with 800 square feet of prime lab space, complete with a dozen benches, adjoining office, and Faraday cage. "I hear he's having those kids build circuit boards. What the hell does he think this is, a *vo-tech school*?" But I was fortunate to have three faculty members solidly on my side (the only ones who had worked in industry), and was buffered by a well-publicized program that gave students the chance to design and build real systems, not just run simulations as an antidote to the textbook world of lumped constants and zero-rise time pulses (*much .edu about nothing*).

The experience did cure me of any lingering wistful regret about having missed those idyllic college years, despite the daily consumption of epicurean cafeteria fare. All around me, students were undergoing a sort of extended agony of uncertainty and stress, many of them deep into engineering school without any notion of what real engineers actually *do*. But what mattered most was assembling an effective student team every quarter, learning how to be a manager, and getting on with the project. My job was to identify and attract folks on the fringe, give them interesting design challenges, and leverage their work to propel myself into the next adventure.

Within a few months we had about ten FORTH nodes scattered around the lab on benches arranged like finger piers in a marina, all chained to a Hub and forming a classroom-scale Microship emulator by a braided 6-wire bus and running a protocol designed by one of our wizard volunteers (giving us pier-to-pier networking).

Nodes were assigned to any task that could be clearly defined, and some of the student teams produced decent work. Audio, serial, and video crossbar networks flickered to life, collectively forming *Grand Central Station*. A couple of shy fellows managed with a little coaching to decode the communications between a Sony DAT recorder and its remote control, then replicate it to allow software to drive the unit. ("This is the happiest moment of my life," one said as they arrived, newly confident, to give their final report in suits and ties.) After a couple of failed attempts, we found two guys to design the controller for the boat's video turret with its steerable high-quality camera: Working with a wizard student from the Mechanical Engineering department who designed the mechanics, they conjured a beautiful piece of work that survived all the technological attrition of the ensuing years and became part of the final system. These were the exceptions, and they made it all worthwhile.

Other projects weren't so successful. The first attempt to write a GUI for the crossbars took two guys two quarters, yielding a fat listing of impenetrable C code for a '286 that talked directly to the display drivers. It never really worked, but the fault lay with my vague specifications; exasperated one afternoon, I learned HyperTalk and wrote a Mac-based front end in about six hours. Some teams never got past their initial learning curves, and one pair working on power management turned their final report into an explanation of the FET, complete with characteristic curves and massive amounts of replicated library material, counting on the ancient tradition of BS to carry the day. One fried the fuse in my beautiful new Fluke multimeter trying to measure the current capacity of a power supply by measuring across it; I sent him downtown on a quest for a replacement with the explanation that one of the first requirements of engineering is knowing how to track down parts. Someone power-cycled my Tektronix scope with every measurement; someone else cut a live multi-conductor power cable with diagonal cutters and made the sparks fly; yet another wired a fully populated DB-25 with about a half-inch stripping on each wire, rendering it virtually untouchable without causing a short. And I'll never forget, "Steve, do these colors on the resistors really mean anything?"

But you know, it was worth it. I relate all this to give a realistic view of the Gaussian distribution curve, which has to be taken into account when accepting free help in a classroom setting. In sum, I consider it a win-win.

Making the Academic Connection

Now that I have either teased you or frightened you with visions of naive students poking with sharpened sticks at your delicate labor of love, I should add that this is not an easy relationship to initiate, nor is it always something you want. Only a small number of engineering schools offer the quality of education that is likely to return more value than it costs you to deploy... if you want to be a volunteer teacher, that's fine, but if you are hoping a team of bright-eyed young geeklets will speed your dream on its way to reality, you'll need to be willing to take charge of a few things. Choose the institution carefully, be sure you are able to drive the project-definition and ongoing assessment processes, maintain close faculty alliances, and put serious time into fine-tuning all the associated relationships.

One more caveat: student project energy is viewed as a scarce resource in the very places where the quality is likely to be highest, and I have run into places where a savvy faculty member is already guarding it jealously with no shortage of pet projects in reserve. Such people will make it highly difficult for you to get a foot in the door, necessitating a higher level of institution-hacking (or, more likely, looking elsewhere entirely).

If your nearby school looks promising and goes well beyond textbooks (or wants to, and sees you as a resource to that end), spend some time making connections among students, professors, and alumni. *What do they want?* PR? IP rights to anything invented on-site? Leveraged corporate sponsorship with the companies you already have on board? Curriculum development? Or just a lighter workload for their own lazy or inexperienced teachers? The latter can make for easy entry, but can also backfire if it is an indicator of overall academic sloppiness... whether the result of incompetent faculty, or tenured people more interested in publication and funding than undergraduate engineering education. This is more common than one might hope, alas.

Chapter 7

What is Gonzo Engineering?

I have casually sprinkled this phrase throughout the preceding pages, knowing that the meaning will gradually emerge from context. But it is worthwhile now to wrap this up by crystallizing this concept... it is central to the the very notion of reaching escape velocity with the combined forces of invention and the tools we have been discussing.

If you probe the interstices of an industry increasingly dominated by Big Business, you'll discover a microculture of hackers motivated by the mad bliss of invention, surviving on the sweet contagion of creative energy. Employment bonuses mean nothing here; fancy packaging and market share are viewed with contempt if a product lacks art. *Beauty*, now that's the thing—the beauty of elegant code, of a robust network, of a balanced design that "just works" without duct tape and feature bloat.

It is from this culture that the Internet emerged, as well as the Open Source movement. Less obviously, it's also a diverse community of home-shop machinists, Arduino artists, guerilla solar experimenters, human-powered vehicle designers, robotics hobbyists, amateur radio satellite builders, and countless other independent developers. If you want to see passionate invention without the sloppy overhead of a big R&D budget or the weird constraints of maximizing shareholder value, go find a hacker... someone who gets a techno-boner from circumventing limitations and knows how to get things done.

This has been my world for 30 years—a world where fun is the bottom line and livings are made on the opportunistic spinoffs of creativity, not selling one's life for a salary. We subsist in the dark matter between industries, trolling flea markets and dumpsters for Obtainium, mail-ordering goodies, making holy pilgrimages to the surplus Mecca of Silicon Valley, re-purposing the detritus of corporate America to our own obsessive ends. Scattered among us are conjurers, alchemists, wizards, lone-wolf inventors, quirky entrepreneurs, larger-than-life writers, and the origins of more than a few disturbing geek stereotypes.

In this parallel universe, the motivation for creating is highly personal. In industry, you can bet that any massive development effort is associated with a business plan—there's no room for slack in a bottom-line world, and seldom are things done for *fun*. But here, you'll find entire lifetimes given over to chasing quixotic dreams; you'll see personal fortunes whittled down to marginal subsistence in the name of invention and reputation. Occasionally there's an imagined pot o' gold, to be sure, but most likely it's just a reassuring fiction to keep the *spousal unit* calm in the face of demonic focus, Every Goddamn Night Out There in the Shed. No, our motives are usually as guileless as passion itself: chasing daydreams, building tools, realizing obsessions, shattering limits, publishing, earning grins of appreciation from the cognoscenti and accolades from neophytes.

These are things that touch the soul more than the bank account, and there's definitely a conceit about it—our sense of security lies more in our toolsets than our 401-Ks. We feel sorry for vested employees with their BMWs and well-appointed houses, even as we decorate our labs with rusted hand-me-down office furniture and pay for system upgrades by mining our hardware boneyards through eBay. But money is not the point. It's the exhilaration of surfing the knee of the learning curve, the almost erotic bliss of a machine flickering to life—catching the spark and glowing while the rest of the world sleeps.

Of course, getting to that point can involve a ludicrous amount of work.

7.1 The First Steps

This is an almost embarrassingly intimate look at how crazy unbalanced people can take an ambitious dream and pull together the resources to make it come true (and then go out and play). You'll never get a corporate middle manager to admit it, but such lunacy, driven by emotion and other unquantifiable wild cards of the psyche, lies at the very heart of the design process. You can formalize tools and implement procedures all you like, but you can't fit passion on a PERT chart; trying to do so will repel the very people you need most.

The first step is one of the most fun: indulging in a fantasy rich enough to trigger secret grins of hard-core technolust. That's the stuff that makes otherwise sensible engineers willing to devote years, if that's what it takes, to getting it right.

One of the great secrets I've discovered is that even someone with stupendously bad work habits (like me) can get a prodigious amount accomplished by applying one simple and obvious technique: *keep moving in the same direction for a long time.* Unfortunately, that can lead one down the path of specialization—an essential part of the great symbiosis between those who dream and those who produce. Specialization along with its concomitant skills is obviously necessary to get real work done, but if you're not careful it can also become a filter through which you see the world, attenuating everything that is not somehow related to your primary focus. Over time, this can cause severe perceptual distortion from which it can be difficult to recover (especially if said specialty ends up, not necessarily through any fault of your own, becoming an evolutionary dead end in a rapidly changing industry).

That's an easy platitude for a self-proclaimed generalist to spout, but how do we resolve the problem? How do we hold on to a central design objective for a decade or more without becoming like one of those single-issue political or religious zealots who lose the broader context entirely and descend into extremism? It's much easier to end up there than you might think, especially when you audaciously choose to chase a personal obsession rather than sell 40-hour weeks while hanging onto the remainder for your own sanity-preserving pursuits.

The trick is at once simple and fiendishly tricky: all it takes is caring so passionately about the project that it fills your daydreams, turns trade journals into treasure hunts, induces you to recruit your friends, inspires doodles, and overlays a sense of purpose onto every foray into the backwaters of the web. This is a lot to ask of a job that's been dumped on you by management, and one of our central messages here is that if this crazy-talk of passion gets you all fired up and chafing at the bonds of a career that isn't letting you play enough, then maybe some restructuring is in order. For there is simply no way that crank-turning, even by a well-oiled department full of Really Smart People, is going to give you a sustained rush of intense creative obsession; doing that requires a suite of characteristics that are generally regarded as pathological in a corporate environment:

- ☐ Enough chutzpah to believe that you are doing something original and important, but the humility to steal shamelessly from the work of those who have preceded you
- ☐ Enough schmoozing ability to induce others to buy in to the dream, but the stubbornness to continue believing in your mad quest when associates have given up on you
- ☐ Enough optimistic naiveté to interpret catastrophic failures as steps along a continuous path, but the sensitivity to recognize the real gotchas (like your own change of heart) when they subtly appear
- ☐ Enough arrogance to ignore the warnings and skepticism of people with far more experience, but the wisdom to shut up and listen quietly to the advice of practitioners in a completely unrelated field

People who behave this way are often described as having attitude problems, difficulty working well with others, and a tendency to jump around and not finish assignments. These are not the things managers look for in employees.

What I'm trying to tell you here is that if you are one of these troublesome folks, you need to shape your environment to support your passions: nothing is more important than removing the barriers that our culture erects around creative madmen, and few companies are willing to customize a job description to allow your brain to go berserk in its own juices. In severe cases, you might even need to jump ship and accept the insecurities that accompany working alone. (On the other hand, if you are in management and are trying to pull off the impossible, then you need to recognize and encourage the hackers in your midst, giving them the freedom to be profoundly annoying and unpredictable.)

All this is simply a contextual backdrop for the real point here, which is that massively audacious feats of creativity fall out of a way of thinking that is much more a *lifestyle* than a *toolset*. I find myself smirking at books about management and team-building, when virtually every world-changing cusp in the fabric of technology can be at least partly attributed to the obsessive-compulsive behavior of some intense character who broke the rules, dropped out of school, irritated colleagues, jumped between careers, got in trouble, or, as the schoolbooks used to say about the inventors I tended to identify with, "died alone in poverty, an embittered man."

It seems we keep returning to this theme: a lifestyle of dedication to a mad dream, with everything else shoved aside as necessary to make room for equipment, learning curves, relationships with gurus and assistants, testing phases, and the endless quest for support. It's not necessarily profitable, nor is it particularly fun (in the amusing sense), but there is something blissful about having a *raison d'etre*, a central passion, an unwavering navigational objective that allows every instant of your life to be tagged unambiguously with Distance To Go, Cross-Track Error, Estimated Time of Arrival, and Speed Over Ground. Such clarity may be illusory, but it beats floundering around every day, changing direction on a whim, and questioning your purpose even while working your butt off and looking forward mostly to evenings, weekends, vacations, and retirement.

It's also no guarantee of success. But even going spectacularly down the tubes feels kind of noble when it's part of your life's enduring quest.

Still, I keep wanting to overlay some kind of formality on this. Aren't there a few rules we can apply that are a bit more useful than saying "just dream it," like some incongruously successful relic of the 60s who became a crystal-sucker in the New Age fringes of Silicon Valley before stumbling into a founder's pool during the can't-fail dot-com boom? *Like, it's all about the fundamental vibrations of your creative energy, man...*

Well, um, yes. But if this level of design is indeed a lifestyle, then the closest we can get to "formal tools" is a body of behaviors and attitudes. Let's put on an engineering hat and attempt to consider the problem in that light.

7.2 Formal Tools, Briefly Considered

Sometimes I wish I could claim that Microship development had been a tightly managed progression in which, beginning with a vaporous initial concept, we generated increasingly refined formal specification documents, mapped everything onto a PERT chart to establish dependencies, used that to drive human resources and purchasing departments, then underwent a tightly scheduled fabrication and coding phase focused on milestones and design reviews. That's how big companies claim to do it... and, hey, we even have some nifty project-management software that knows how to convert TO-DO lists into pretty pictures.

During the *BEHEMOTH* era, I spent a very interesting afternoon at Trimble Navigation, makers of the bike's GPS. These weren't colorful, user-friendly handhelds wrapped around off-the-shelf chipsets back then; they were extremely complex DSP engines coupled with RF hybrid black magic that pushed just about every envelope in the book. I remember being captivated by a massive floor-to-ceiling PERT chart, spanning an entire hallway, the completed boxes bright yellow, the web of interconnections revealing Deep Understanding of the design process and accurate predictions of every step remaining.

"I should do this for the bike," I mused to my host. "It looks like a great tool."

"Nah," he replied. "Project management tools assign resources to tasks. You work alone. Just do *something*."

He was right. Even with first-class volunteers and occasional contract help, Nomadic Research Labs is a tiny operation, a de facto non-profit, beset by overload and bad work habits, constantly challenged by such fundamental issues as demotivation, distraction, and lack of funds. A PERT chart in this environment would be masturbatory, and would presuppose a stable design.

7.3 Engineering in a Nutshell

What actually happened was much more organic, and I've noted with amusement that, despite protestations to the contrary among the engineering population, it's typical of the way things usually work in industry. Here's how to manage a huge, complex project:

1. Accept going in that your first tentative decomposition of the fundamental concept will yield an over-simplified TO-DO list, distorted by misunderstanding of key issues.
2. Avoiding all the items labeled TBDWL (To Be Dealt With Later) or ATAMO (And Then A Miracle Occurs), dive headlong into the well-defined parts, finishing some of the electronic design so early in the game that it is guaranteed to be obsolete before the physical substrate is built.
3. Blunder ahead on the non-obvious parts, getting pleasantly distracted by learning curves and occasional moments of certainty, only to discover basic flaws in your reasoning.
4. Now that you are forced to re-think the initial concept, map it onto newly recognized reality to yield a fresh TO-DO list (with new lab notebooks and computational tools to keep things lively) and another cycle of enthusiastic activity.
5. Repeat steps 3-4 countless times at varying levels of abstraction ranging from the entire system down to individual components.
6. Meanwhile, since technology evolves with frightening rapidity, acknowledge the fact that any computer-based system is such a moving target that if it's not completed quickly, it will be irrelevant by the time it ships.
7. Respond by simplifying the design, further refining your objectives and abandoning dead-end ideas while doggedly pursuing others that have come to

represent too large an economic or emotional investment to allow a graceful retreat.

8. Compromise here and there, bang out a few things that weren't on the list, then add them and cross them off to make yourself feel good.

9. Get totally sidetracked a few times, and periodically dive into major development marathons to meet public deadlines like trade shows, pulling all-nighters in PFD mode (Procrastination Followed by Despair).

10. Announce new completion dates whenever a previously predicted one has passed, and keep driving your PR engine to maintain interest during a process that is a textbook illustration of Hofstadter's Law *("Everything takes longer than you expect, even when you take into account Hofstadter's law.")*

Part of this development heuristic is just sloppy management, but it also reflects the way we think. This is why engineering is, at its heart, an art form (and why the average completion time of a homebuilt boat is 135 years).

Perhaps the most interesting thing about this seemingly ugly process is that it's iterative and self-correcting. Grandiose or stupid ideas may not be obvious during first-pass blue-sky analysis (when the project is glued together by wishful thinking), but it's another story entirely when it all has to be converted into Clearly-Defined Tasks (CDTs) and drawings that make sense to machinists. Without some kind of closed-loop intellectual process to fine-tune your thinking, it would be impossible to get to the point where you can start using engineering tools to convert fantasies into contraptions.

Trying to shortcut this by starting on Day One with formal design methodologies can have the catastrophic effect of committing you to an ill-defined goal state, whereupon the end result is shaped more by your toolkit than by the supposed objective. That's why so many products seem malformed, patched, and otherwise inelegant: management loves formal methods and looks askance upon such frivolous notions as approaching product design as a delicate blend of art and engineering. The exceptions, when they occur, are a joy to use. The rest miss the point, no matter how stylish their exterior or sophisticated their underlying technology.

So it appears that designing a system isn't nearly as rigid a process as typical engineering textbooks would have you believe. Your component choices affect the shape of the thing you're building; said shape in turn creates constraints that affect your choice of components. Such psychological race conditions can only be resolved by tweaking the granularity knob while adding inputs to your evolving mental model, until the correct solution congeals in a flash.

> **It's easy, and here's how to do it: Prop your feet up on your desk, relax, and form a fantasy of the desired results. Now turn it slowly in your head while calmly examining it from all sides, allowing input variables to float until an unanticipated combination satisfies your psychic fantasy-comparator and generates a flash of recognition. Since all your noodling is saved in a big circular buffer called short-term memory, let this recognition event pre-trigger a snapshot of the conditions that immediately preceded it (before accumulated pondering-propagation delays introduce conceptual drift). There's your design specification. Take that and run with it!**

This is probably not an engineering methodology that makes managers comfortable, though it's a good summary of life in the trenches. There is a pervasive myth that structured methods and sequential procedures, used in isolation, will get you there... but I've never seen it work that way. The tools don't actually start to become useful until you're quite thoroughly immersed, and that can take weeks of appearing, to outside observers, as if you are loafing.

7.4 A Sense of Urgency

Speaking of time, there's another big difference between gonzo engineering and life in industry. Schedules and deadlines, the X-axis of project management, are anathema to the independent worker. Don't tell me that I have until Monday morning at 9:00 to hand you a report on the solar array thermal retrofit; I'm still in the wall-staring phase on that one and expect to be here for days! I might emerge occasionally to troll the web for prior art to steal, get distracted by other parts of the project, or just say "screw it" and go sailing on a friend's catamaran in the name of research, but a deadline? Imposing order on the project would send me on a search for something better suited to my interpretation of the term "work."

Alas, life isn't like that in a corporate environment, where people actually pay you to behave. Critical-path management, release dates, pre-production prototypes, purchasing cycles, trade shows... there are countless reasons why the long-suffering denizens of cubicles and labs are not given free rein to go with their instincts. But despite the importance of scheduling in coordinating a complex enterprise, there are huge costs involved: design compromises, sneaky shortcuts, employee burnout, kluged patches, bad assumptions, useless documentation, and incomplete testing, just to name a few. This is analogous to sailing: it is well understood that a sailor with no schedule always has fair winds. The people who find themselves calling MAYDAY in a Force 10 gale are usually those who have decided to push their luck for some time-related reason: they're in a race, vacation's almost over, the crew has to reach port in time to use a return ticket, or some arbitrary schedule laid out over charts and cruising guides in a cozy den long ago is now affecting the skipper's judgment.

Working alone and with volunteers on something that will be done when it's done (and not before), we have the luxury of ignoring the calendar—although with that comes the dangerous temptation to give in to the dreaded BEHEMOTH Effect ("Hey, here's a cool gadget; let's see how we can integrate it into the system!") Somewhere in there is the right compromise, but we are going to assume that when you're building your life around the Ultimate Project, schedules are not a factor.

Convenient, eh?

7.5 Economics

Finally, let's talk about money. From an engineering perspective, this can be even more annoying than time — there's nothing like "aggressive cost minimization" to take all the fun out of a design. Fortunately, one of the intrinsic features of passionate dream-chasing is that everything else is secondary, and it's thus easy to justify spending as much as you have (and then some). Combine this with poverty consciousness, and one can get amazingly creative at scrounging. In addition to all the expensive bits from West Marine and McMaster-Carr, our crazy research vessel contains thousands of parts that were donated, bought surplus, extracted from dumpsters, horse-traded, repurposed, cannibalized, or fabricated on the cheap. But one issue that never came up was worrying about manufacturability and component cost. There's a sort of certainty here that is immensely liberating:

"This is the most important thing I can possibly be doing, so it doesn't really matter what it costs to get the job done. I'll afford it somehow."

Afterword

I hope I have given you some useful tools in this little publication. Every gonzo engineering project depends most obviously upon whatever flavor of geek expressionism lies at its core, yet it is excruciatingly vulnerable to the realities of human limitation. The basic rule is to just keep moving in the same direction for a long time, but that isn't enough... even if you are of the Wizard class, eventually you still need stuff, exposure, cash flow, and help.

The tips I have presented here are not theoretical. In 1983 I fled the suburban lifestyle and took off across the US on a recumbent bicycle. 17,000 miles later, I was on the third version of the bike and had a book under my belt as well as relentless media coverage, a team of volunteers, and enough corporate sponsors to make the massive undertaking possible. A crazy dream had turned into a career; now, over 25 years later, I am still reaping the benefits of this process when working on my new projects.

I mentioned at the beginning that I was about to reveal all my "trade secrets," things I've never publicly discussed before. This is true, and the reason is that it is kind of a zero-sum game. In the heat of the *BEHEMOTH* bicycle era, I was not only careful about inviting competition into my odd little niche, but also conscious of the delicacy involved in the intertwined mesh of relationships... all critical to the ongoing operation. It would not have felt right to publicly explain how to schmooze for goodies or gently manipulate the media into helping build a consistent image when I was dependent upon such relationships for my very existence.

But this is valuable information, as timely as ever, and as I make the transition to a full-time life aboard a voyaging sailboat I find that I'm willing to share now. Enjoy. And good luck!

If you want to follow my adventures, the blog at http://www.nomadness.com/blog and the site front door at http://microship.com will get you started. I'm also on Twitter as *nomadness*.

Cheers!
Steve

www.ingramcontent.com/pod-product-compliance
Lightning Source LLC
LaVergne TN
LVHW010107110826
845155LV00028B/530

* 9 7 8 1 9 2 9 4 7 0 0 5 1 *